WELDING

FOR
ARTS AND CRAFTS

WELDING

FOR
ARTS AND CRAFTS

Dewayne Roy

THOMSON

DELMAR LEARNING

Australia   Canada   Mexico   Singapore   Spain   United Kingdom   United States

Welding for Arts and Crafts

Dewayne Roy

Executive Director:

Alar Elken

Executive Editor:

Sandy Clark

Acquisitions Editor:

Sanjeev Rao

Development Editor:

Alison Weintraub

Executive Marketing Manager:

Maura Theriault

Channel Manager:

Fair Huntoon

Marketing Coordinator:

Brian McGrath

Executive Production Manager:

Mary Ellen Black

Production Editor:

Ruth Fisher

Art/Design Coordinator:

Cheri Plasse

Editorial Assistant:

Jill Carnahan

Library of Congress Cataloging-in-Publication Data

Dewayne, Roy
  Welding for arts and crafts / Dewayne Roy.
    p. cm.
  ISBN 0-7668-1896-9 (alk. paper)
  1. Welding. I. Title.

TT211.D48 2000
739'.14—dc21   99-089024

This book is dedicated to my parents who have given me the past, my wife who is always by my side, my son to whom I give the future, and my dog Oscar who has been at my side for the writing of this book from start to finish. I love you all.

CONTENTS

PREFACE

Think of this book as a tool for dreamers as well as doers. Its intent is not to instruct welding or fabrication but rather to put a spark in your vision, inspiration, and creative fantasies. It's a place to get fresh ideas and new designs. It's all about doing new, unexpected, and stylish things with welding.

Hopefully, you will find that the welded art projects that you make will create a specific mood or style that is functional and aesthetically pleasing. Let the projects be a guideline for your own creativity. The materials they are made from can easily be changed to suit your own taste. As in every project in the book, be creative and have fun.

TRANSFERRING PATTERNS

The author has furnished you with specific patterns for each project. To enlarge or reduce a pattern, simply use an office quality copy machine, cut out the pattern with scissors, and trace the pattern with soapstone by placing it on a sheet of metal and marking completely around it. If your copy machine does not enlarge, contact a copy shop. Having a copy made is usually inexpensive and can be made quickly. Check the yellow pages of your local phone directory under "copying" or "photocopying" for the shop nearest you.

Note: Patterns are not to scale.

CHAPTER

1

WELDING

Welding is the fusing of two or more pieces of metal to form one. It is a precise, reliable, cost-effective method for joining metals. Today, designs and creations in metal are often combined with different welding processes to create a product. No one specific rule controls the selection of a welding process for any given job. There are more than 80 different welding processes used in industry today.

The primary goal of this chapter is to give you the basic information on the advantages and disadvantages of the welding processes used to build the projects in this book.

SHIELDED METAL ARC WELDING

Shielded Metal Arc Welding (SMAW), also known as stick, arc, and electric welding, is the most common form of arc welding. See Figure 1-1. It is one of the most versatile ways to weld since the filler material or welding electrode can easily be changed to match different metals just by switching the electrodes. This process can be used to weld aluminum, carbon steel, cast iron, nickel, and stainless steel.

Advantages

SMAW is often used because of its versatility. When compared to the other types of power sources, the SMA welding machines are less expensive and more portable. This process is very flexible in terms of the metal thickness that can be welded. Metal as thin as 1/16 inch (2 mm) thick or several feet thick can be welded using the same machine, but with different settings. As a result, the SMAW process is often utilized by novice welders, farmers, and fabrication and maintenance shops to weld on a variety of jobs.

FIGURE 1-1   *SMAW machines are relatively inexpensive and very popular for home workshops.*

Disadvantages

The amount of skill required by the operator is greater than that needed by the other processes. The amount of downtime that is associated with this process can be a deterrent if it is used for production. The electrode is only so many inches in length and must be changed as it is consumed. This requires the user to stop welding and change the electrode. Another disadvantage is that all welds must be either chipped or ground to remove the slag from the weld.

GAS TUNGSTEN ARC WELDING

GTAW, also known as Tungsten Inert Gas (TIG) or Heliarc welding, is the most common process for welding aluminum. See Figure 1-2. The GTAW process produces high quality welds on almost all metals and alloys. Because it can be easily controlled at very low amperages, the GTAW process is ideally suited for welding on sheet metals; however, it tends to be an expensive process to use for welding metals thicker than 1/4 inch (6 mm) because of the speed of welding.

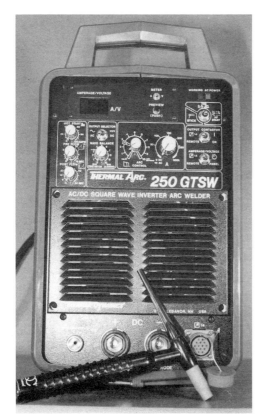

FIGURE 1-2 *The GTAW machines date back to the 1930s. They are ideal for welding aluminum, magnesium, stainless steel, and carbon steel.*

Advantages

The GTAW process produces high quality welds on almost any weldable metal or alloy. Because fluxes are not used, the welds produced are sound, free of contaminants and slags, and are as corrosion-resistant as the parent metal. The GTAW process is used because of the high-quality welds that it makes and for the good appearance of the welds.

Disadvantages

The welder must coordinate precise movement of the torch with one hand, while adding filler rod with the other hand, and then controlling the current with a foot pedal.

The setup of the equipment is more complicated than the other processes. The welding operator must have an understanding of tungsten preparation and current selection, choosing whether to weld AC or DC, and if welding DC, whether to weld using DC+ or DC–. It is also helpful to understand spark intensity, high frequency, upslope, downslope, background current, pulse rate, peak intensity, and proper grounding.

GAS METAL ARC WELDING

Gas Metal Arc Welding (GMAW), also known as Metal Inert Gas (MIG), wire, or semiautomatic, is a process that uses a spool of wire. The spool or wire is either housed inside a power source or fed through an external wire feeder into a welding gun. See Figure 1-3.

Advantages

The GMAW process does not require the degree of welding operator skill that the GTAW or SMAW processes do. The welding operator sets the current on the welding machine, holds the gun in one hand, squeezes the trigger, and welds. It's that easy. GMAW is a fast-growing process, in part, because compact welding units now retail for less than $550 and the process provides the ability to easily weld on thinner metals compared to the SMAW process. GMA welding can be used on both thick and thin sections of metal. It can be used to weld steel, aluminum, stainless steel, and many other metals. For these reasons, this process is the choice of many fabrication and production shops.

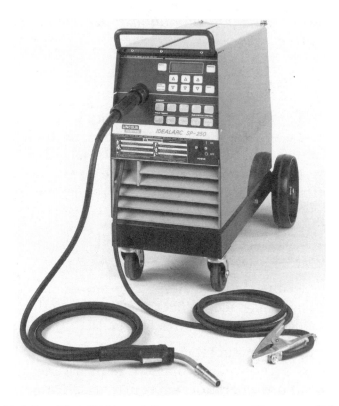

FIGURE 1-3 *The GMA welding process utilizes a solid metal wire and requires the use of a shielding gas to protect the weld pool from the atmosphere.*

Disadvantages

The GMA process is less portable than the SMAW equipment, which means that the work must be brought to a welding station; portable models do exist. Welding outside with this process is somewhat troublesome; if the wind blows the shielding gas away, it produces inferior welds. More maintenance is required of this process because of the associated moving parts.

PLASMA ARC CUTTING

Plasma is the state of matter that is found in the region of an electrical arc. The plasma created by an arc is an ionized gas that has both electrons and positive ions whose charges are nearly equal to each other. See Figure 1-4.

Advantages

High travel speeds can be used with plasma cutting, which eliminates most distortion particularly on the sections of metal. All metals can be plasma cut. Most plasma cutting machines built today are easy to operate.

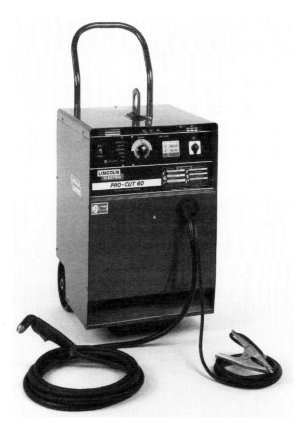

FIGURE 1-4   *Any material that is electrically conductive can be cut using the plasma arc cutting process.*

Disadvantages

The open circuit voltage is higher for Plasma Arc Cutting than for the other processes; extra caution must be taken. Water is sometimes used to cool the plasma torches, which improves the cutting characteristics. Any time water is used, it is very important that there be no leaks or splashes. The chance of electrical shock is greatly increased if moisture is on the cables, floor, or equipment.

Plasma cutting is noisy. Ear protection is required to prevent auditory damage to the operator or others working in the area.

The plasma cutting process produces light radiation in all three spectrums. Proper eye protection is a must when using this process.

This process produces a large quantity of fumes that are potentially hazardous; a means for removing them from the work space is needed.

Consumable parts of the torch must be replaced as they wear out or become damaged. This can become very expensive if the equipment is not operating correctly.

OXYFUEL CUTTING

Oxyfuel Gas Cutting (OFC) is a process that uses a high-temperature oxyfuel gas flame to preheat the metal to a kindling temperature, at which time it will react rapidly with a stream of pure oxygen. See Figure 1-5. The most common gases used with this process are acetylene, hydrogen, **m**ethyl**a**cetylene **prop**adiene (MAPP), natural gas, propane, and propylene.

Advantages

Attachments can be added to the torch for heating, cutting, welding, and brazing.

The overall cost of the equipment is low, in comparison to other welding or cutting equipment. The portability of the equipment is superb since you do not need a power source to operate it. The skill level is quick to achieve; however, OFC is unfortunately one of the most commonly misused processes. Most workers know how to light the torch and make a cut, but their cuts are very poor quality and often unsafe. A good oxyfuel cut should not only be straight and square, but it should require little or no cleanup.

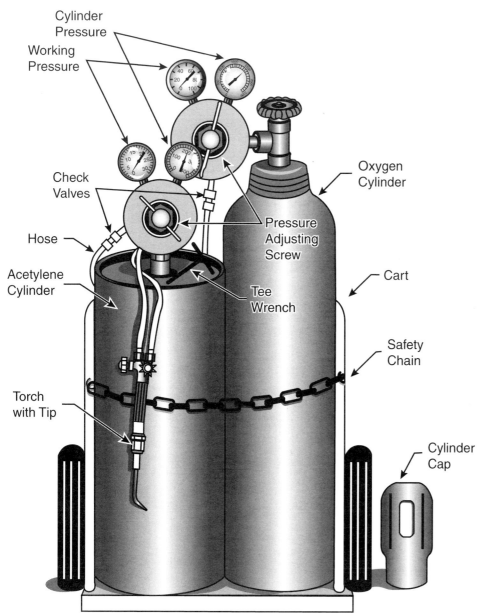

FIGURE 1-5 *The oxyacetylene gas cutting torch is the most commonly used oxyfuel gas cutting torch in the industry.*

Disadvantages

The OFC process is primarily used for cutting steel. It will not cut stainless, aluminum, copper, or brass. If a metal will not rust, it will not cut properly with this process.

SMAW TROUBLESHOOTING GUIDE

Welding Problem	Remedy
Arc Blow	1. Adjust the electrode angle. 2. Move the ground clamp. 3. Use AC current instead of DC. 4. Inspect the part to see if it has become magnetized and if so, demagnetize.
Cracks in Weld	1. Reduce the welding speed, use an electrode that has a more convex bead. 2. Use low-hydrogen electrodes. 3. Use pre- and post- heat procedures on the weld.
Distortion	1. Reduce the amount of current. 2. Use chill plates. 3. Increase your welding speed. 4. Clamp or fix the parts being welded. 5. Weld thick sections first.
Porosity	1. Use dry electrodes. 2. Do not weld on wet metal. 3. Clean paint, grease, oil, and dirt from the metal being welded. 4. Shorten the arc length. 5. Use low-hydrogen electrodes.
Slag Inclusions	1. Increase the amount of current. 2. Decrease your welding speed. 3. Do not allow the welding pool to get ahead of the arc. 4. Change the polarities. 5. If welding multi-pass welds, always chip and wire brush between passes.
Spatter	1. Decrease the amount of current. 2. Shorten the arc length. 3. Weld on dry metal with dry electrodes.
Undercutting	1. Decrease the amount of current. 2. Reduce the welding speed. 3. Shorten the arc length. 4. Change the electrode angle.

GTAW TROUBLESHOOTING GUIDE

Welding Problem	Remedy
Arc Blow	1. Make sure the ground is properly connected. 2. Change the angle of the electrode relative to the work to stabilize the arc. 3. Make sure the base metal is clean. 4. Make sure the tungsten electrode is clean. 5. Use a smaller tungsten electrode. 6. Bring the nozzle closer to the work.

Welding Problem	Remedy
Brittle Welds	1. Check the type of filler metal being used. 2. Use pre- and post- heat procedures while welding. 3. Check the welding procedure. A. Multilayer welds will tend to anneal, hard heat affected zone.

Cracked Welds	1. Check for inferior welds. Make sure all welds are sound and the fusion is good.
	2. Check the type of filler metal being used.
	3. Check the welding procedure.
	   A. Fill all craters at the end of the weld passes.
	   B. Use pre- and post- heat procedures.
	4. Do not use too small a weld between heavy plates.
	5. Check for proper preparation of the joints.
	6. Check for excessive rigidity of the joints.

Inferior Weld Appearance	1. Change the welding technique. 2. Use the proper amount of current. 3. Check for proper shielding gas. 4. Check for drafts blowing the shielding gas away. 5. Check for proper filler metal. 6. Make sure the base metal is clean.
Distortion	1. Make sure the parts are properly tack welded together. 2. Clamp the parts to resist shrinkage. 3. Remove the rolling or forming strains by stress relieving techniques before welding. 4. Distribute the welding to prevent excessive local heating.
Porosity	1. Check the shielding gas. 2. Check for drafts blowing the shielding gas away. 3. Do not weld on wet metal. 4. Make sure the base metal is clean. 5. Check the shielding gas hoses; plastic hoses are best. 6. Check the welding procedure and current settings.

Rapid Tungsten Electrode Consumption	1. Check the polarity.
	2. Decrease the amount of current.
	3. Use a larger tungsten electrode.
	4. Check the shielding gas.
	   A. Increase the gas flow.
	5. Check for good collet contact.
	   A. Change the collet.
	   B. Use ground-finished tungsten.

Tungsten Inclusions	1. Use a larger tungsten electrode or reduce the amount of current. 2. Do not allow the tungsten to come into contact with the molten weld pool while welding. 3. Use a high-frequency starting device. 4. Do not weld with tungsten electrodes that have become cracked or split on the ends.

(continued)

GTAW TROUBLESHOOTING GUIDE *(continued)*

Welding Problem	Remedy
Lack of Penetration	1. Increase the amount of current. 2. Decrease your welding speed. 3. Decrease the size of the filler metal.
Undercutting	1. Decrease the amount of current. 2. Make smaller weld beads. 3. Use a closer tungsten-to-work distance. 4. Change your welding technique.

GMAW TROUBLESHOOTING GUIDE

Welding Problem	Remedy
Arc Blow	1. Change the gun angle. 2. Move the ground clamp. 3. Use backup bars made of brass or copper. 4. Demagnetize the part.
Cracked Welds	1. Check the filler wire compatibility with the base metal. 2. Use pre- and post-heat procedures on the weldment. 3. Use a convex weld bead. 4. Check the design of the root opening. 5. Change the welding speed. 6. Change the shielding gas.
Dirty Welds	1. Decrease the gun angle. 2. Hold the gun nozzle closer to work. 3. Increase the gas flow. 4. Clean the weld joint area and gas flow. 5. Check for drafts that may be blowing shielding gas away. 6. Check the gun nozzle for damaged or worn parts. 7. Center the contact tip in the gun nozzle. 8. Clean the filler wire before it enters the wire drive. 9. Check the cables and gun for air or water leaks. 10. Keep unused filler wire in shipping containers.
Wide Weld Bead	1. Increase your welding speed. 2. Reduce the amount of current. 3. Use a different welding technique. 4. Shorten the arc length.
Incomplete Penetration	1. Increase the amount of current. 2. Reduce your welding speed. 3. Shorten the arc length. 4. Increase the root opening. 5. Change the gun angle.

Welding Problem	Remedy
Irregular Arc Start	1. Use wire cutters to cut off the end of the filler wire before starting a new weld. 2. Check the ground. 3. Check the contact tip. 4. Check the polarity. 5. Check for drafts. 6. Increase the gas flow.
Irregular Wirefeed Burn-back	1. Check the contact tip. 2. Check the wire feed speed. 3. Increase the drive roll pressure. 4. Check the voltage. 5. Check the polarity. 6. Check the wire spool for kinks or bends. 7. Clean or replace the worn conduit liner.
Welding Cables Overheating	1. Check for loose cable connections. 2. Use larger cables. 3. Use shorter cables. 4. Decrease the welding time.
Porosity	1. Check for drafts. 2. Check the shielding gas. 3. Increase the gas flow. 4. Decrease the gun angle. 5. Hold the nozzle close to the work. 6. Do not weld if the metal is wet. 7. Clean the weld joint area. 8. Center the contact tip with the gun nozzle. 9. Check the gun nozzle for damage. 10. Check the gun and cables for air or water leaks.
Spatter	1. Change the gun angle. 2. Shorten the arc length. 3. Decrease the wire speed. 4. Check for drafts.
Undercutting	1. Reduce the current. 2. Change the gun angle. 3. Use a different welding technique. 4. Reduce the welding speed. 5. Shorten the arc length.
Incomplete Fusion	1. Increase the current. 2. Change the welding technique. 3. Shorten the arc length. 4. Check the joint preparation. 5. Clean the weld joint area.
Unstable Arc	1. Clean the weld area. 2. Check the contact tip. 3. Check for loose cable connections.

CHAPTER

2

WELDING SAFETY

Practice safety at all times! Safety is not to be taken lightly when using welding equipment. Welding is as safe or as hazardous as you make it; the decision is up to you and you alone.

Accident prevention is the primary intent of this chapter. Many welding, cutting, and allied processes can be harmful to the health of any untrained person. The author of this book is a recognized welding instructor, with many years of experience teaching welding technology at the college level. He makes it mandatory for his students to understand the dangers of welding before they are allowed to enter the welding laboratory. The rules that the students use are as follows:

ROY'S RULES FOR WELDING

- Never weld without proper eye protection.

  Safety glasses are to be worn in the shop at all times.

  Welding helmets with the correct shade of lens are to be used when doing any type of arc welding.

  Welding goggles with the correct shade of lens are to be used when using the oxyfuel process.

- Wear proper clothing.

  Long-sleeved shirts are to be worn while welding. Wool clothing (100% wool) is the best choice, but they are expensive and hard to find. One hundred percent cotton is a good second choice. Synthetic materials such as polyester, dacron, rayon, and nylon should be avoided because they burn easily. The light from the welding processes can cause burns to any skin that may be exposed.

  All-leather boots should be worn while welding.

Pants with legs long enough to cover the tops of the boot are a must and should be without cuffs, holes, or frayed edges.

All-leather gauntlet-type gloves are to be worn while welding.

- Do not breathe welding fumes.

Welding on some materials may produce fumes and gases hazardous to your health.

Keep your head out of the fumes.

Use ventilation to exhaust fumes and gases from your breathing zone and general area.

- Remove all flammable materials such as paper, oil, and cloth from the welding vicinity.

- Have a suitable fire extinguisher conveniently located at all times while welding.

- Do not weld in a building with wooden floors unless the floors are protected from hot metal.

- Earplugs or earmuffs should be worn while welding. The welding environment can be noisy. Hot sparks can fly into an unprotected ear causing severe burns.

- Do not weld on gas tanks, oil or gas drums, or any other type of vessel with a large volume and small opening such as a fifty-five gallon drum barrel.

- Before welding or cutting, warn those in close proximity who are not protected by proper clothing.

- Do not weld in the presence of water.

- The workpiece being welded and the frame or chassis of all electrically powered machines must be connected to a good ground.

- Have all installation, operation, maintenance, and repair work performed only by qualified people.

- Do not wrap cables carrying welding current around your body.

- Do not grind aluminum, copper, brass, or any other nonferrous metal on a conventional grinding stone. If a ferrous stone is used to grind nonferrous metals, the stone will become glazed (metal clogs the surface) and may explode due to frictional heat buildup.

- Oxygen and fuel gas cylinders must be stored separately, and the storage area must be separated by twenty feet.

- Cylinders must be secured with a chain or other device so they cannot be knocked over accidentally.

- Cylinders not in use must have a valve protection cap. The protection cap prevents the valve from being broken off if the cylinder is knocked over.

- Acetylene cylinders must be used in an upright position, never in a horizontal position.

- Oil or grease in the presence of oxygen can cause explosions. Keep all oil-based compounds away from all oxygen cylinders, regulators, and hoses.

- Never use acetylene at pressures in excess of 15 psi.

- The acetylene valve on the cylinder should never be opened more than one and a half turns.

- The oxygen valve on the cylinder should be opened all the way.

- Do not test for fuel leaks with an open flame; test with a leak-detecting solution only.

- Light torches with spark lighters—never with matches or cigarette lighters.

- Butane lighters and matches may catch fire or explode if they are subjected to welding heat or sparks. There is no safe place to carry these items when welding.

- Do not weld in places where dust or other combustible particles are suspended in air or where explosive vapors are present.

- Do not permit unauthorized persons to use welding or cutting equipment.

SAFETY ORGANIZATIONS AND ASSOCIATIONS

American Conference of Governmental Industrial Hygienists (ACGIH) publications. *Threshold Limit Values for Chemical Substances and Physical Agents in the Workroom Environment,* available from American Conference of Government Industrial Hygienists, 1330 Kemper Meadow Drive, Cincinnati, OH 45240.

American National Standards Institute (ANSI). *Safety in Welding, Cutting, and Allied Processes, Z49.1,* available from American Welding Society, 550 N.W. Lejeune Road, Miami, FL 33126.

American Welding Society (AWS). *Study of Fumes and Gases in the Welding Environment,* and other safety and health publications available from American Welding Society, 550 N.W. Lejeune Road, Miami, FL 33126.

National Fire Protection Association. *Cutting and Welding Processes, NFPA Standard 51B,* available from National Fire Protection Association, Batterymarch Park, Quincy, MA 02269.

Occupational Safety and Health Administration (OSHA). *Code of Federal Regulations, Title 29, Labor, Chapter XVII, Parts 1901.1 to 1910.1450,* Order No. 869-019-00111-5, available from Superintendent of Documents, U.S. Government Printing Office, Washington, DC 20402.

For specific information, refer to the Applicable Material Safety Data Sheet (MSDS) available from the manufacturer, distributor, or supplier.

3

YARD ART

CHAPTER

Think of yard art as "Jewelry for the Garden." Imagine an army of metal ants marching through the dwarf maples, a school of fish swimming through the daisies, a rusty cow standing in a patch of bull nettles, an angel hanging on a fence holding a basket of flowering devil's head, tarnished frogs lazily sitting on stones beside the lily pond or fairies hovering over the narcissus.

Finding things to use as yard art is an active pursuit, not an armchair diversion. You cannot create awe-inspiring yard art without getting your hands dirty, and that is part of the fun.

For the projects in this chapter, forget convention and use flea market, salvage yard, or garage sale finds. These will range from old shovels, horseshoes, lightning rods, plow discs, clothes hangers, brass doorknobs, or old water faucets. This chapter will help you turn these ordinary-looking or out-of-date objects into interesting focal points for your yard.

Note: Patterns are not to scale.

PROJECT 1: GARDEN SHOVEL ANGEL

Hang her from a fence, or place a basket of flowers in her hands and watch her weather beautifully.

Supply List

1 old shovel (round or square point)

2 clothes hangers

1 1/2″ diameter flat washer

1 horseshoe

1 piece of 1/8″ thick steel 4″ wide × 12″ long

1 piece of corrugated steel 2′ × 2′ (this is the type of steel used for barn roofs.)

1 piece of 1/4″ steel rod 2′ long

Equipment Needed

a SMA welding machine

a welding hood

a pair of safety glasses

a pair of leather gloves

an oxyfuel torch

a pair of cutting goggles

a spark lighter

a soapstone

vise grip pliers

Patterns

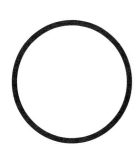

Circle

Hand

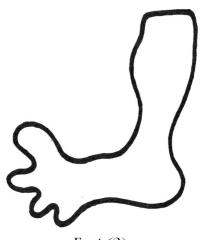

Foot (2)

Wing (2)

STEP 1-1 - *Cut the horseshoe in half.*

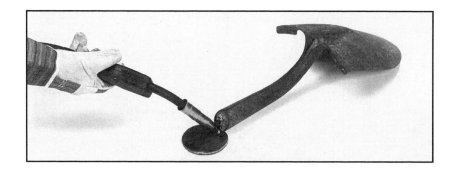

STEP 1-2 - *Make a 3/4-inch long weld on the back side of the 4-inch circle, affixing it to the shovel.*

STEP 1-3 - *Connect the wings and washer to the back of the shovel with a series of small spot welds.*

STEP 1-4 - *Weld the arms (half horseshoes) in place and affix the hand to the arms, making sure that the thumb is pointing upwards toward the head.*

STEP 1-5 - *Place the feet at the bottom of the shovel on the front side, but weld from the back.*

STEP 1-6 - *Make the ring of the halo by bending the 1/4-inch rod completely around a 4-inch pipe.*

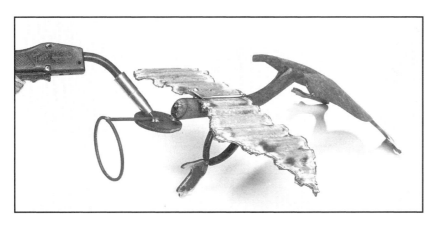

STEP 1-7 - *Weld the 1/4-inch circle portion of the halo to a 1/4-inch × 6-inch rod and affix it to the back of the head.*

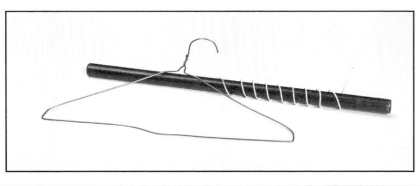

STEP 1-8 - *Straighten out the clothes hangers and bend coils around a 1-inch pipe.*

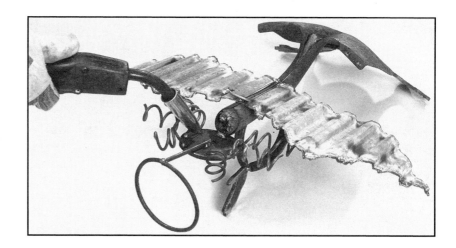

STEP 1-9 - *Cut coils into six equal pieces and weld to the back of the head.*

PROJECT 2: GARDEN FAIRY

This is an easy project if you are proficient with an oxyfuel cutting torch. You will probably want to make several, because they will look great hovering over the flowers in your garden. You may also want to paint her if you wish or place her in the garden as she is.

Supply List

1 piece of 1/4" thick steel 6" wide × 12" long

1 piece of 16 gauge steel 4" wide × 10" long

1 piece of 1/4" steel rod × 36" long

Equipment Needed

an oxyfuel cutting torch

a pair of cutting goggles

a spark lighter

a pair of safety glasses

a 4 1/2" angle grinder

a GMA welding machine

a welding hood

a pair of leather gloves

a soapstone

a plasma arc cutting machine

Patterns

Wing (2)

Fairy

STEP 2-1 - *Clean up all rough edges with the grinder and affix the wings to the fairy body by welding.* Note: *The wings should be tilted outward away from the body at a slight angle.*

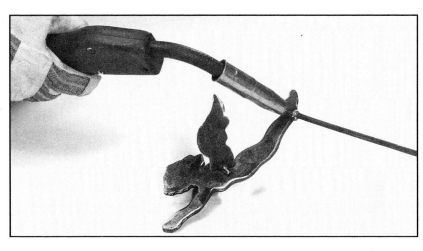

STEP 2-2 - *Weld the 1/4-inch rod to the feet.*

PROJECT 3: BIRDBATH

When the project is complete, put the birdbath in your yard, and you'll have the cleanest birds around.

Supply List

1 plow disc (roughly 20″ in diameter and the deeper the better). The plow discs can be purchased from local farm supply stores.

1 finial (a finial is the decorative piece that is often used for the top of a steel picket fence).

3 30 ″ long 1/2″ twisted rod (*see the sources at the back of the book for a list of companies that sell this type of product*).

1 10″ circle (3/8″ thick)

4 steel balls (1″ in diameter)

3 7″ rings made from 1/2″ square tubing (*see the sources at the back of the book for a list of companies that sell this type of product*).

Equipment Needed

a GMA welding machine

a welding hood

a plasma arc cutting machine

a pair of safety glasses

a pair of cutting goggles

a pair of leather gloves

a soapstone

a torpedo level

Patterns

Large leaves (3)

Small leaves (3)

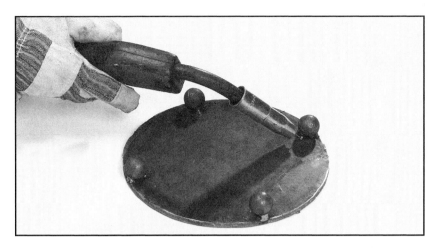

STEP 3-1 - *When working with the 10-inch circle, think of it as a clock and weld the 1-inch steel balls at 12-3-6-9 o'clock positions.*

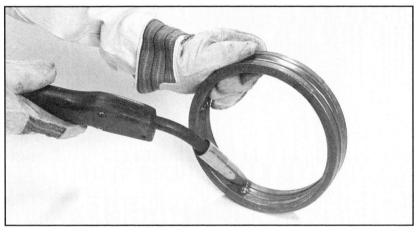

STEP 3-2 - *Stack and align the 3-inch to 7-inch rings and spot weld them together on the inside surfaces. Place four spot welds 1/2-inch long at 12-3-6-9 o'clock positions on each joint.*

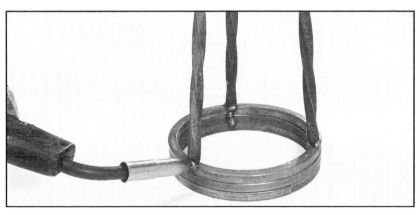

STEP 3-3 - *Tack weld the three twisted rods to the 7-inch rings 120 degrees apart from each other.*

STEP 3-4 - *Since the twisted rods are only tack welded, move the opposite ends of the three rods together until they touch each other, then tack weld the points that come into contact with each other.*

STEP 3-5 - *Affix the three twisted rods that have been tacked together to the center of the 10-inch circle. Tack weld them to the plate that serves as the base.*

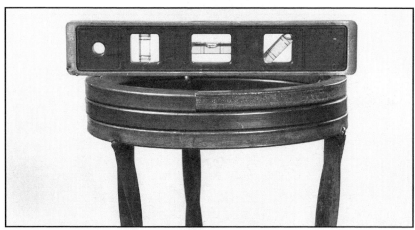

STEP 3-6 - *Place your torpedo level on top of the three rings. Check it for level both north-south and east-west, making sure that everything is level. Now weld the 1/2-inch twisted rod all the way around in all the places that you previously tacked.*

STEP 3-7 - *Now weld the finial in the center of the plow disc. Weld from the bottom if possible. Remember, the weld needs to be watertight.*

STEP 3-8 - *Place, center, and level the plow disc on top of the rings and place a 1-inch weld in 12-3-6-9 o'clock positions.*

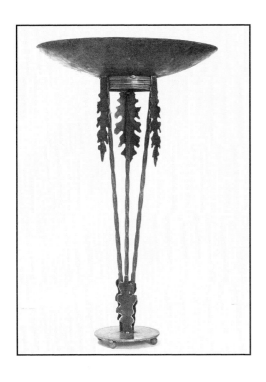

STEP 3-9 - *Place the leaves on the birdbath and spot weld each to the tops of the twisted rods.*

PROJECT 4: WATER HOSE HOLDER

This next project will work equally well with contemporary or classic homes, because of its clean lines. The water hose holder will add character and charm to your garden—plus it's easy to do.

Supply List

1 decorative cast-iron design approximate size (7″ wide by 15″ long). (If you are unable to find one from either a salvage yard or flea market, *see the sources at the back of the book for a list of companies that sell this type of product.*)

1 5/8″ twisted bar 5′ long

1 finial for 5/8″ bar

1 1/2″ square tubing 14″ long

1 3/4″ steel ball

1 1/4″ × 1″ × 5″ flat steel bar

Equipment Needed

a GMA welding machine

a welding hood

a pair of leather gloves

a pair of safety glasses

a tape measure

a soapstone

STEP 4-1 - *Weld the decorative cast iron design 9 inches from one end to the 3/4-inch twisted base.*

STEP 4-2 - *Weld the 5/8-inch finial to the top of the 5/8-inch twisted bar.*

STEP 4-3 - *Bend the 1/2-inch square tubing around an 8-inch piece of pipe until you have a 90-degree bend.*

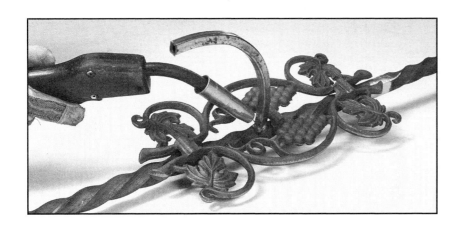

STEP 4-4 - *Weld the 90-degree bend to the 5/8-inch twisted back. Do not weld to the casting because the casting is not structurally sound. It is decorative only.*

STEP 4-5 - *Weld the 3/4-inch steel ball to the end of the 1/2-inch tubing.*

STEP 4-6 - *Center and weld the 1/4-inch × 1-inch flat bar to the 5/8-inch twisted bar 18 inches from the bottom. This flat bar will enable you to place your foot on the hose holder to push it into the ground.*

PROJECT 5: FLOWER CART

Enhance the beauty of those potted plants around the house with this next project. Put a flower cart on the patio or at the front door to create a memorable first impression that can turn any unassuming doorway or patio into an enchanting entryway to the garden.

Supply List

12 feet of 1/8″ × 1″ steel flat bar

10 feet of 1″ square tubing 1/8″ wall thickness steel

27 inches of 1/2″ square tubing steel

1 horseshoe

1 3/4″ nut

50 inches of 1/4″ steel rod

2 pieces of 1/4″ × 1″ × 16″ steel flat bar

2 of the same decorative cast iron designs (approximate size 4″ wide by 29″ long) (*See the sources at the back of the book for a list of companies that sell this type of product.*)

Equipment Needed

a chop saw

a speed square

a tape measure

a GMA welding machine

a welding hood

a torpedo level

a pair of leather gloves

a soapstone

a pair of safety glasses

STEP 5-1 - *Using the chop saw, cut two pieces of the 1-inch square tubing 29-inches long and cut two pieces 6-inches long. Take the four pieces, make a rectangle, and tack them together. Use the speed square to assure that the pieces are square.*

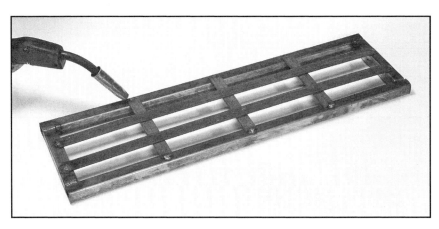

STEP 5-2 - *Using the chop saw, cut three pieces of the 1/8-inch × 1-inch flat bar 7 inches long and cut two pieces 28-inches long. Uniformly space them on the bottom of the rectangle and weld them in place.*

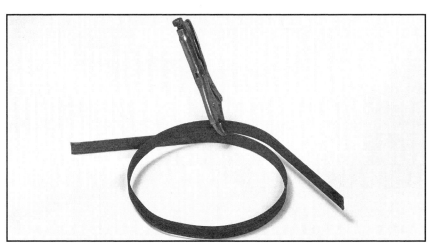

STEP 5-3 - *Using the chop saw, cut one piece of 1/8-inch × 1-inch steel flat bar 55 1/2-inches long and bend it around a section of 10-inch pipe to form a circle. You will have approximately two feet of waste, but this is necessary in order to get the leverage you need for bending.*

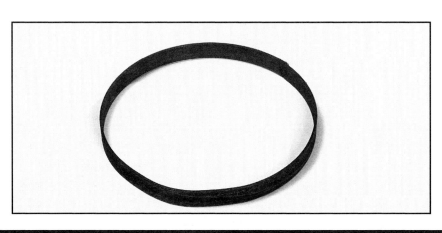

STEP 5-4 - *Again using the chop saw, cut off any excess so that you have formed a circle and then weld the seam together.*

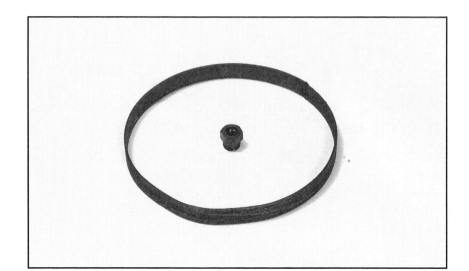

STEP 5-5 - *Place the circle on a flat surface, centering the 3/4-inch nut.*

STEP 5-6 - *Using a tape measure, measure from each point of the nut to the inside edge of the circle and cut a piece of 1/4-inch rod to length. It is necessary to do this for each point because the circle is most likely not a perfect circle. Weld the nut to the rod and the rod to the circle. You have now completed the steel wheel. Since the design of the wheel is strictly decorative, not functional, the wheel does not have to be perfect.*

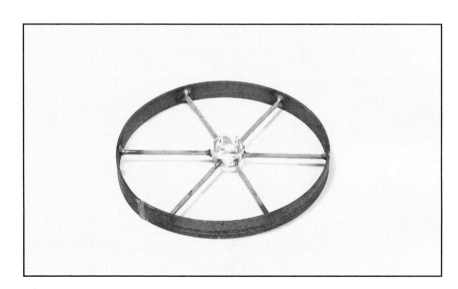

STEP 5-7 - *Tack weld each 16-inch piece of 1/4-inch × 1-inch steel bar to each side of the 3/4-inch nut. Then tack weld the steel to the other end of the steel bar and to the center of the rectangle.*

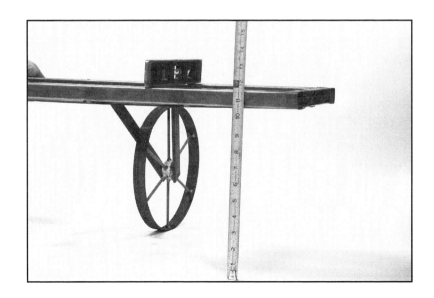

STEP 5-8 - *Using the torpedo level, level the rectangular frame and get a measurement from the bottom of the rectangular frame to the floor. Using that measurement, cut two pieces of 1-inch square tubing and tack weld to the rectangular frame to make legs.*

STEP 5-9 - *Place the 27-inch piece of 1/2-inch tubing in the center of the two upright 1-inch square tubing legs and tack weld into place. Make sure it is square with the legs before you continue.*

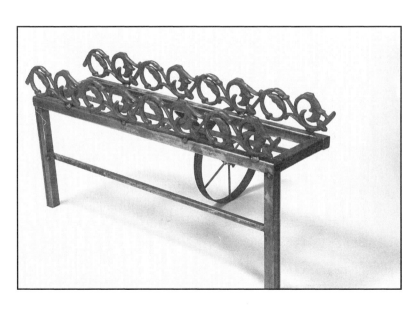

STEP 5-10 - *Weld the two 29-inch decorative cast iron designs to both the front and back of the rectangle.*

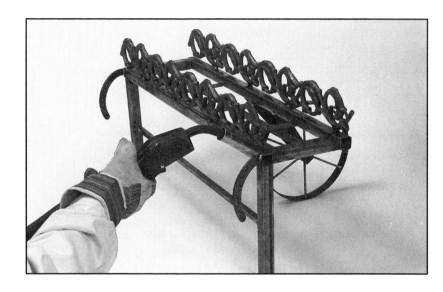

STEP 5-11 - *Cut the horseshoe in half and weld each half to either back side corner of the cart as handles.*

Weld all joints that had previously been tack welded; if the cart is square, then all legs should touch the floor; if not, make necessary adjustments.

PROJECT 6: DECORATIVE PLANT HOLDER

This is a project that any gardener would love, and the size of the plant holder can easily be altered to your specific need. It can be hung from a tree limb or from a metal hook in the ceiling. Experiment by placing handmade metal leaves from your region in the holder or add scrolls to the plant holder to create a unique look. Have fun.

Supply List

11 pieces of 1/8" × 1" steel flat bar 9" long

4 pieces of 1/4" steel rod 14" long

4 pieces of 1/4" steel O.D. 9" long

1 piece of 1/4" steel rod 12" long

1 piece of 16 gauge steel 16" × 16"

1 piece of 1/2" square tube 1" long

1 finial that fits 1/2" square tubing

Equipment Needed

a GMA welding machine

a welding hood

a pair of safety glasses

vise grip pliers

a pair of leather gloves

a plasma arc cutting machine

a pair of cutting goggles

a speed square

a soapstone

Patterns

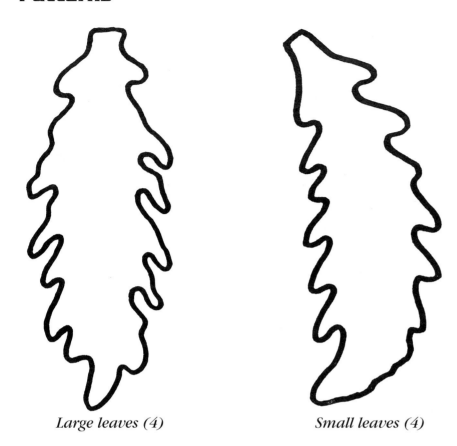

Large leaves (4)   *Small leaves (4)*

STEP 6-1 - *Using eight of the 9-inch long pieces of 1/8-inch × 1-inch steel bars, make two squares. Be sure to tack weld the outside of the squares.*

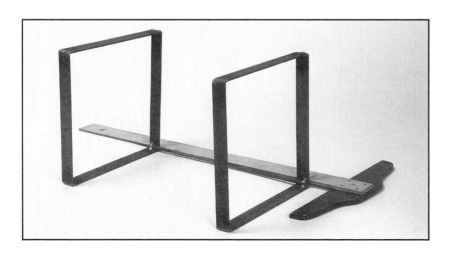

STEP 6-2 - *Using the straight edge, place the squares on their sides 12 inches apart. This makes the outside edges 14 inches apart. Place one of the 14-inch steel rods in the corner of the two squares and tack weld. Repeat this procedure to all four corners.*

STEP 6-3 - *On top of the 14-inch steel rods, tack weld the 9-inch long, 1/4-inch steel rods in the corners of the box.*

STEP 6-4 - *Weld the finial onto the 1-inch long, 1/2-inch square tube and affix it to the four 1/4-inch rods at the point of contact.*

STEP 6-5 - *At the end of the box opposite the location of the finial, space three of the 1/8-inch × 1-inch × 9-inch flat bars evenly and weld them into place. At this time, you can perform permanent welds on all the places that had previously been only tack welded.*

STEP 6-6 - *Weld the large decorative leaves on each of the four top corners of the plant holder.*

STEP 6-7 - *Weld two of the smaller leaves on both of the front and back of the bottom of the plant holder.*

STEP 6-8 - *Using the 12-inch long piece of the 1/4-inch steel rod, bend it into a circle approximately four inches in diameter and weld it to the 1/2-inch square tube surrounding the finial on top of the plant holder.*

PROJECT 7: RAIN GAUGE

Amidst the hustle and bustle of our daily lives, we all need a place for serenity. Hopefully, this next project with its clean distinctive lines and oriental character, will add some of the serenity to your garden found in the gardens of China and Japan.

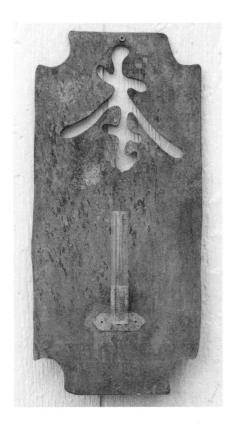

Supply List

1 piece of 1/4″ thick steel 4″ wide × 12″ long

1 glass vial and holder from a rain gauge. (This can be bought very inexpensively from a hardware store.)

Equipment Needed

a GTA welding machine

a welding hood

a pair of safety glasses

a pair of leather gloves

an oxyfuel cutting torch

a soapstone

a pair of cutting goggles

a sparklighter

a 4 1/2″ angle grinder

a 1/4″ drill motor with 1/4″ drill bit

Pattern

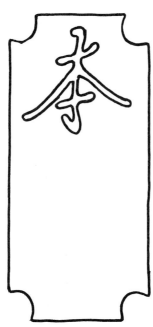

STEP 7-1 - *Using a correctly lit torch, cut out the pattern. Clean any ragged edges with the 4 1/2-inch angle grinder.*

STEP 7-2 - *Using the drill motor and 1/4-inch drill bit, drill a hole at both the top and bottom of the 1/4-inch plate. The placement of the holes may vary depending on where the gauge is to be placed.*

STEP 7-3 - *Remove the glass vial from its holder and place the holder in the center of the 1/4-inch plate one inch from the bottom of the rain gauge. Using the GTA welding machine, place a small weld on the left and right sides of the holder.*

When your project is complete, place the rain gauge in your garden in an upright position and wait for the next rain shower.

CHAPTER 4

SOMETHING FISHY

Mention fish and most people think of eating or catching one, but how many people have thought of fish as an art form?

Someone once said "Give me a fish and I'll eat today, teach me to fish and I'll eat for a lifetime." Hopefully, the fish in this chapter will bring to your mind's eye enough images to expand on for the rest of your life.

The designs in this chapter are intended as a guideline for your own creativity. The materials the designs are made from can easily be changed to suit your own taste. As in every project in this book, be creative and have fun.

Note: Patterns are not to scale.

PROJECT 1: PICTURE FRAME

Everyone has a friend or someone in the family who loves to fish. This project makes a great gift for that person, and they will love it because you made it.

Supply List

1/4″ × 12″ × 18″ carbon steel plate

3″ × 5″ piece of glass

10″ of 1/2″ angle iron 1/8″ thick

a piece of cardboard 1/8″ thick × 3″ wide × 5″ long

Equipment Needed

an oxyfuel cutting torch

a pair of safety glasses

a pair of cutting goggles

a spark lighter

a soapstone

a GMA welding machine

a 4 1/2″ angle grinder

Patterns

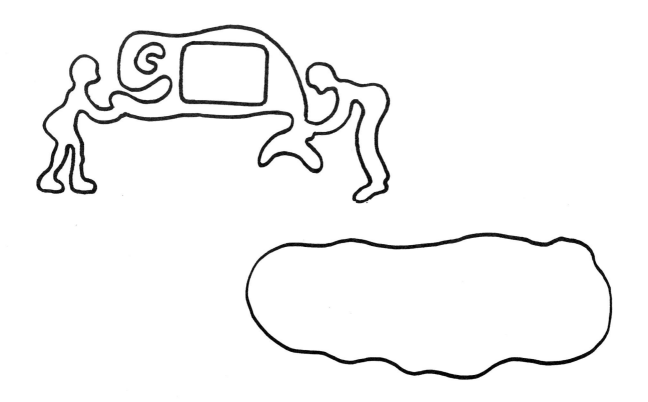

STEP 1-1 - *Using the soapstone, draw a 2 1/2-inch × 4 1/2-inch rectangle onto the body of the fish and cut it out using the cutting torch. If there are any rough edges, smooth them with the 4 1/2-inch angle grinder.*

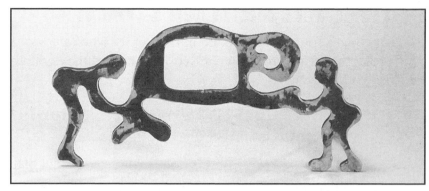

STEP 1-2 - *Cut two pieces of 1/2-inch angle iron 4 1/2-inches long and weld one piece to the top and bottom of the rectangle 3 1/4 inches apart and facing each other to make brackets to hold the glass and picture.*

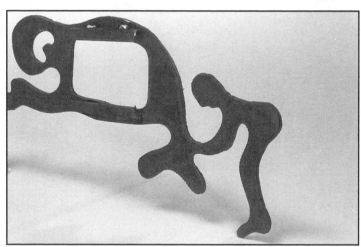

STEP 1-3 - *When working on the second pattern, smooth off any rough edges and weld it to the main structure to make a base.*

STEP 1-4 - *Make sure that the 3-inch × 5-inch piece of glass and matching piece of cardboard slide easily into the 1/2-inch angle iron. If they do not, make any needed adjustments.*

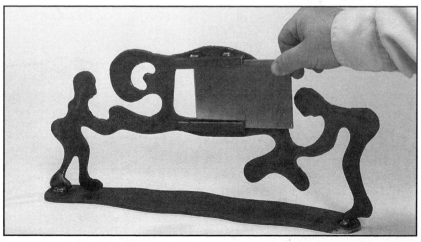

PROJECT 2: WHALE OF A TAIL

Supply List

1/2″ × 8″ × 12″ carbon steel plate

a piece of 1/4″ steel rod 1″ long

a base for a tail 4″ × 4″ (it can be made from stone, metal, or wood); the choice is yours

Equipment Needed

an oxyfuel cutting torch

a spark lighter

a pair of cutting goggles

a 4 1/2″ grinder

a GMA welding machine

a shop vise

a 12″ adjustable wrench

Pattern

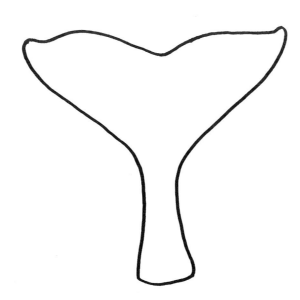

STEP 2-1 - *After cutting out the pattern, use the 4 1/2-inch angle grinder to round all the edges on the tail and taper the end of the trail on both sides. The taper length is approximately 1 1/2-inches long.*

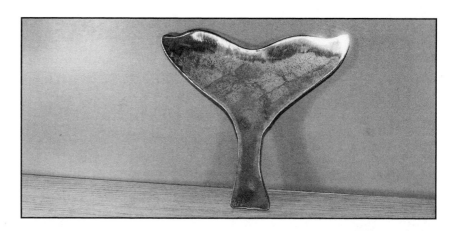

STEP 2-2 - *Place the tail into the jaws of the shop vice with the wide portion of the tail facing upward. Using a torch, heat the portion that is tapered until red hot. Do this all the way across the width of the tail. While the tail is red hot, place the adjustable wrench on the tail and bend it slightly, giving it a slight radius.*

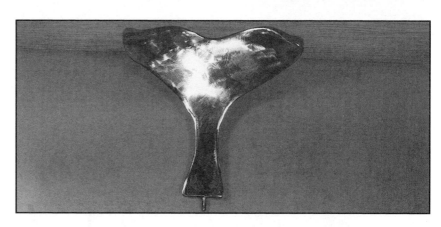

STEP 2-3 - *Weld a 1/4-inch rod in the center of the small end of tail if you are mounting the tail to stone, wood, or glass. Do not use the 1/4-inch rod is using a metal base. Instead, weld the tail directly to the metal base.*

STEP 2-4 - *Using the 4 1/2-inch angle grinder, remove the discoloration from the heating and bending process.*

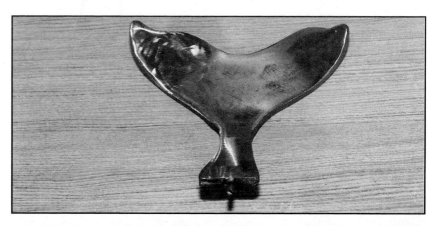

PROJECT 3: FISH LAMP

Accessorize your home, cabin, or office with a fishy flair! This next project is simple and plain, yet unique, and it is rapidly becoming popular with interior designers.

When you have completed the assembly of your project, take it to an electrician or lamp shop to be wired for electrical service.

Supply List

2 pieces of 1/2″ twisted bar 2′ long

1 piece of 1/2″ square tubing 2′ long

1 9″ circle 1/8″ thick

3 solid steel balls 1″ in diameter

1 piece of 16 gauge carbon steel sheet 14″ × 18″

1 piece of 16 gauge copper sheet 6″ × 6″

1 3/8″ nipple and nut 3/4″ long for light socket

1 piece of 1/8″ × 1″ carbon steel flat bar 7″ long

1 piece of brass sheet 8″ × 12″

3 pieces of 1/4″ copper tubing 1 1/2″ long

3 pieces of carbon steel rod 1/8″ diameter × 12″ long

Equipment Needed

a plasma arc cutting machine

a speed square

a soapstone

a pair of cutting goggles

a GMA welding machine

oxyfuel welding torch

vise grip pliers

a pair of cutting goggles

a spark lighter

a 4 1/2″ angle grinder

Patterns

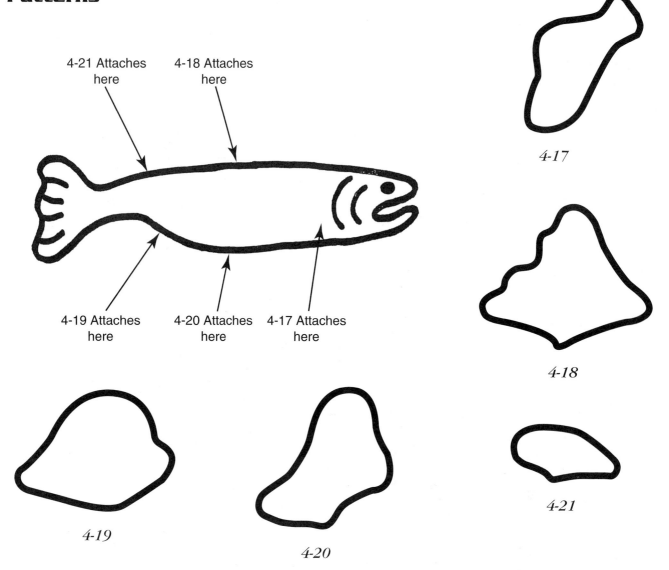

4-21 Attaches here

4-18 Attaches here

4-19 Attaches here

4-20 Attaches here

4-17 Attaches here

4-17

4-18

4-19

4-20

4-21

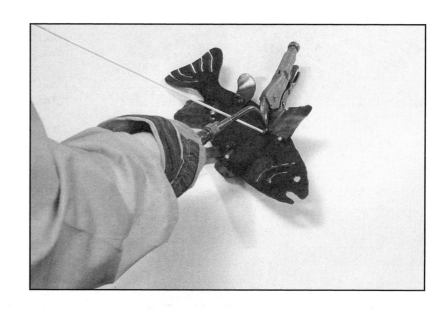

STEP 3-1 - *On both fish, lap the copper fins over the fish body by 1/8 inch, then clamp them securely using the vice-grip pliers. Braze the copper to the steel by using the oxyfuel equipment.*

STEP 3-2 - *After brazing the fins to the body of the fish, remove any flux residue and allow the pieces to cool. Using your hands, bend the bodies of the fish to give them a slight curve.*

STEP 3-3 - *Weld the 1-inch diameter balls to the outside edge of the 9-inch circle 120 degrees in relation to each other.*

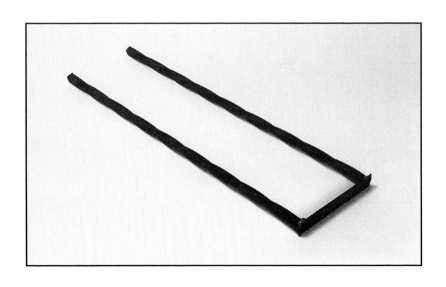

STEP 3-4 - *Weld the 1/8-inch × 1-inch × 7 -inch carbon steel flat bar to the two pieces of 1/2-inch twisted bar. Use the speed square to assure that the three pieces are placed 90 degrees in relation to each other.*

STEP 3-5 - *On the 9-inch circle, center and weld the twisted bars to the sides opposite the 1-inch steel balls.*

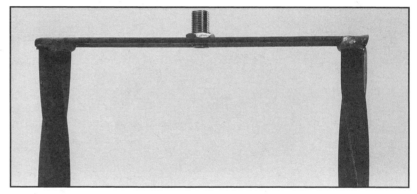

STEP 3-6 - *Using the plasma arc cutting machine, cut a hole large enough for the 3/8-inch nipple to fit into the center of the 1/8-inch × 7-inch flat bar and braze it into place. The nipple should be at a 90 degree angle to the 1/8-inch flat bar. This is what the light bulb socket will later screw into.*

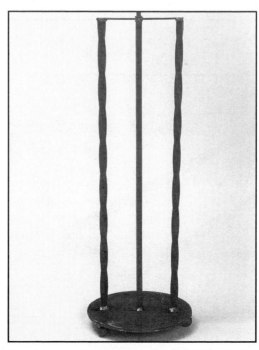

STEP 3-7 - *Align the 2-foot-long, 1/2-inch diameter square tubing with the 3/8-inch nipple, and tack weld it to the 9-inch circle and 1/8-inch flat bar. Make sure the tubing is square with both the 9-inch circle and the 1/2-inch twisted bar.*

STEP 3-8 - *Centering the fish on the twisted bar, weld them from the back side. There is no right or wrong place to position the fish; use your own creativity.*

STEP 3-9 - *Using an oxyfuel welding tip and a slightly oxidized flame, cut the brass sheet into several 8-inch to 12-inch aquatic leaves (narrow and long), and braze them together on one end.*

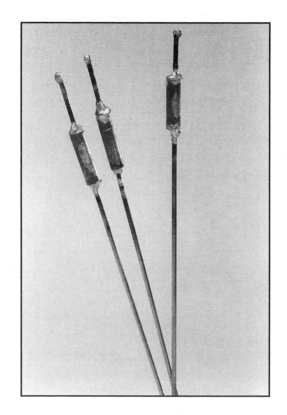

STEP 3-10 - *Slide each piece of 1/4-inch copper tubing onto each piece of 1/8-inch diameter rod approximatley 2 inches from one end, and then braze weld the tubing to the rod. These will be used as cattails in conjunction with the brass aquatic leaves.*

STEP 3-11 - *Braze weld the cattails to the brass leaves and then braze the whole leaf assembly to the 1/2-inch square tubing at the point it joins the 9-inch circle.*

PROJECT 4: A FISH BIRD FEEDER?

Dress up your yard with this next project. One look and you can see that the design is somewhat different from the norm, but the lines are clean and refreshing. One look and you will discover that you can easily modify it to your own taste. You will also discover that this project will last a lifetime and is easy to do.

Now take a look at how to build a fish bird feeder.

Supply List

1/4″ × 12″ × 12″ carbon steel plate

8″ × 8″ carbon steel sheet 1/16″ thick

1 piece of 3/8″ steel rod 48″ long

Equipment Needed

an oxyfuel cutting torch

a pair of safety glasses

a pair of cutting goggles

a spark lighter

a soapstone

a GMA welding machine

a 4 1/2″ angle grinder

a plasma arc cutting machine

Patterns

STEP 4-1 - *Overlap the fish with the man and weld from the back side. Weld the leaf to the man's hand.*

STEP 4-2 - *Weld the 3/8-inch round rod to the back side of the fish's body.*

When you have completed the fish bird feeder, paint it, place in your yard, and wait for the birds.

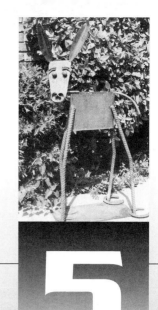

CHAPTER 5

CRITTERS

Oftentimes I have people ask me where I get my ideas for projects. This next chapter deals with that question. Many artists have found inspiration in the things that are around them everyday in order to create their exquisite work.

Open your eyes, look around, and see what you can see.

Note: Patterns are not to scale.

PROJECT 1: THE LITTLE DONKEY

This is an easy project and fun to make.

Supply List

1 piece of 3/8″ steel plate 8″ wide × 10″ long

4 pieces of 3/4″ concrete reinforcing rod 2 feet long

2 pieces of 16 gauge carbon steel 4″ × 10″

5 small horseshoes

2 1″ ball bearings

1 piece 1/4″ × 1″ × 6″ steel bar

1 piece 1/4″″ steel plate 5″ wide × 10″ long

1 piece 1/8″ steel sheet 12″ wide × 18″ long

1 piece 3/8″ concrete reinforcing rod 9″ long

Equipment Needed

a pair of safety glasses

a welding hood

an oxyfuel cutting torch

a pair of cutting goggles

a pair of pliers

a torpedo level

a spark lighter

a pair of leather gloves

a GMA welding machine

a 4 1/2″ angle grinder

a plasma arc cutting machine

Patterns

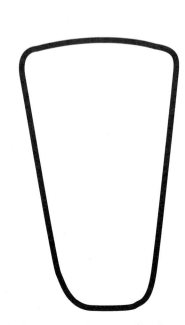

STEP 1-1 - *Using the oxyfuel torch, heat a 2-inch space in the center of each of the 3/4-inch concrete reinforcing rods and bend approximately 30 degrees.*

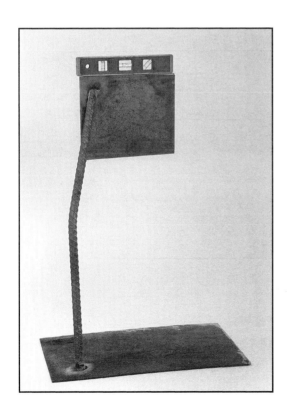

STEP 1-2 - *Weld one of the 3/4-inch reinforcing rods to the front of the 3/8-inch carbon steel and affix the other end of the rod to the 1/8-inch × 12-inch × 18-inch steel sheet. Use the torpedo level to assure that the 3/8-inch plate is plumb and level.*

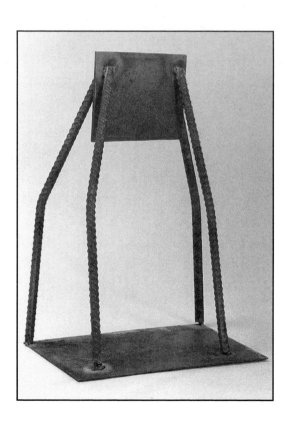

STEP 1-3 - *Weld one of the other 3/4-inch reinforcing rods directly opposite the first one. Notice that the bend of the knee is opposite the first leg. Check once more to make sure the 3/8-inch plate is plumb and level.*

STEP 1-4 - *Weld the remaining 3/4-inch reinforcing rods to the 3/8-inch plate.*

STEP 1-5 - *Bend the piece of 1/4-inch × 1-inch × 6-inch steel bar in the center 90 degrees, and weld it to the 3/8-inch plate.*

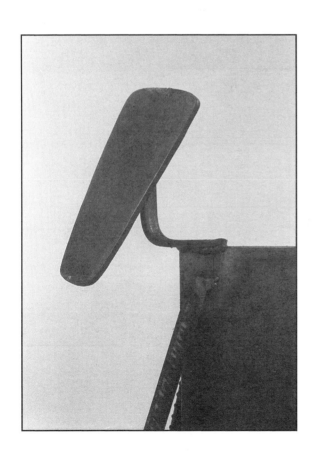

STEP 1-6 - *Weld the head to the 1/4-inch × 1-inch × 6-inch steel bar.*

STEP 1-7 - *Weld each of the ears to the head and bend them slightly forward.*

STEP 1-8 - *Weld the 1-inch balls to the head for eyeballs and use one of the horseshoes for eyebrows.*

STEP 1-9 - *On the opposite end from the head of the 3/8-inch plate, weld the 9-inch piece of 3/8-inch reinforcing rod and bend it slightly to form a tail.*

STEP 1-10 - *Weld the remaining four horseshoes to the bottom of the 3/4-inch reinforcing rod.*

PROJECT 2: LARGE COPPER WINGED DRAGONFLY

The dragonfly is fast and easy to make, looks great around the house, and makes a fantastic gift.

If you place the copper winged dragonfly outside, the wings will turn green in time.

Supply List

4 pieces of 1/2″ copper tubing 6″ long

1 piece of 1/2″ thick steel 1 1/2″ wide × 9″ long

1 1/4″ rod 36″ long

Equipment Needed

an anvil

a hammer

a welding hood

a pair of safety glasses

a pair of leather gloves

an oxyfuel cutting torch

a pair of cutting goggles

a sparklighter

a GMA welding machine

a pair of pliers

a plasma arc cutting machine

Pattern

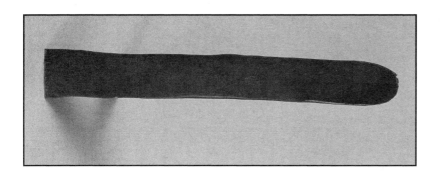

STEP 2-1 - *Hammer each of the 6-inch pieces of copper tubing flat using the anvil and hammer. Round one end of the flattened copper tube with the plasma arc cutting machine.*

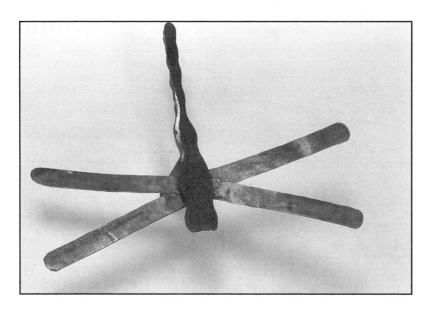

STEP 2-2 - *Braze weld the four flattened pieces to the body of the dragonfly to form wings.*

STEP 2-3 - *Weld the 36-inch long rod to the bottom side of the dragonfly.*

PROJECT 3: SMALL COPPER WINGED DRAGONFLY

Stop! Alto! Halt! Throwing away those old burnt up GMAW contact tips? Finally, there is a use for them.

This next project is even easier to make than the large copper winged dragonfly; it looks great in multiplicity, and it, too, makes a great gift.

Supply List

4 old burnt up GMAW contact tips (hammered flat)

1 horseshoe nail or flat flooring nail

1 1/8″ × 36″ steel rod

Equipment Needed

an anvil

a hammer

an oxyfuel welding equipment

a pair of cutting goggles

a sparklighter

a pair of pliers

a pair of safety glasses

STEP 3-1 - *Place the four flattened contact tips on the horseshoe nail and braze weld.*

STEP 3-2 - *Weld the 1/8-inch × 36-inch steel rod to the bottom of the dragonfly. The dragonflies can be placed outdoors, and the copper wings will eventually turn green.*

PROJECT 4: LEAPING FROG

As long as you have a clean tip and a steady hand, this next project will be a cinch.

Supply List

2 3/8″ flat washers

1 3/8″ diameter steel rod 36″ long

1 piece of 1/4″ copper tubing 8 inches long

1 piece of 1/4″ thick carbon steel 9″ wide × 14″ long

Equipment Needed

an anvil

a hammer

a pair of pliers

a pair of leather gloves

a 4 1/2 angle grinder

a welding hood

a pair of safety glasses

a GTA welding machine

a 1/8″ × 36″ silicon bronze filler metal

an oxyfuel cutting torch

Pattern

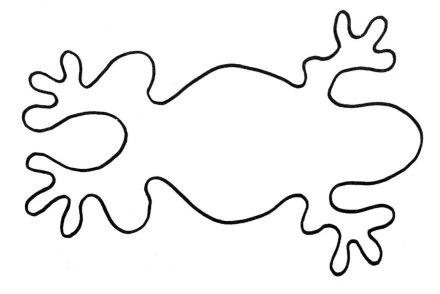

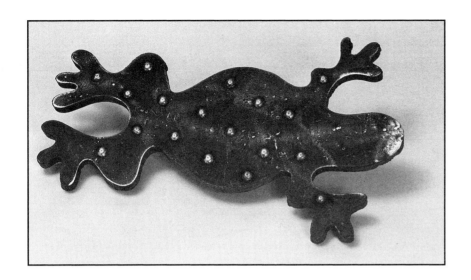

STEP 4-1 - *Using the GTA welding machine and the silicon bronze filler, place spot welds over the entire surface of one side of the frog.*

STEP 4-2 - *Tack weld the two 3/8-inch flat washers onto the frog's head for eyes.*

STEP 4-3 - *Flatten out the 1/4-inch copper tubing on the anvil and weld it on the frog's head for a tongue.*

STEP 4-4 - *Heat the tongue using the oxyfuel equipment until it is red hot. Using the pliers, grab the end of the tongue and curl it towards the head of the frog.*

STEP 4-5 - *Weld the 3/8-inch × 36-inch steel rod to the bottom of the frog.*

PROJECT 5: LOUNGING FROG

This is another easy and fun project. The lounging frog design is somewhat different, because it has clean, crisp lines.

Supply List

1/2″ thick × 12″ × 12″ steel plate

1 ball bearing approximate size 1/4″ diameter

Equipment Needed

a 4 1/2″ angle grinder

a GTA welding machine

an oxyfuel cutting torch

a welding hood

a pair of cutting goggles

a sparklighter

a 1/8″ × 36″ silicon bronze filler metal

a pair of safety glasses

a pair of leather gloves

a 12″ adjustable wrench

a shop vise

a soapstone

Pattern

STEP 5-1 - *Place the frog's body into the jaws of the vise. Use the oxyfuel torch to heat the long back leg of the frog until it becomes red hot; place the adjustable wrench on the end of the leg and bend until you have bent it 90 degrees in relation to the body.*

STEP 5-2 - *Readjust the frog in the jaws of the vise and heat both front legs until they become red hot. Bend the legs 90 degrees in relation to the body.*

STEP 5-3 - *Readjust the frog one more time in the jaws of the vise and heat the remaining leg, but bend it only 45 degrees in relation to the body.*

STEP 5-4 - *Using the GTA welding machine, weld the two 1/4-inch ball bearings on the head for eyes.*

STEP 5-5 - *Using the GTA welding machine and the silicon bronze filler, place spot welds over the entire top surface of the frog.*

PROJECT 6: A BIRDBRAIN PROJECT

This is one project that offers a whole new dimension to the word "birdbrain." Build several of these and integrate them throughout the garden.

All of the components to this project can be purchased from metal architectural suppliers. (*See the sources at the back of the book for a list of companies that sell this type of product.*)

Supply List

1 forged steel basket approximate size 2 1/2″ wide × 5″ long

1 steel ball 1″ in diameter

2 steel wings approximate size 2″ wide × 4 1/2″ long

1 horseshoe nail

1 1/4″ rod 36″ long

Equipment Needed

a GTA welding machine

a welding hood

a pair of leather gloves

a pair of safety glasses

a pair of pliers

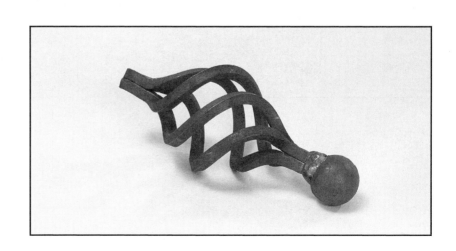

STEP 6-1 - *Using the pliers to hold the steel ball, weld it to one end of the forged steel basket.*

STEP 6-2 - *Using the pliers to hold each of the wings, weld them to the forged steel basket.*

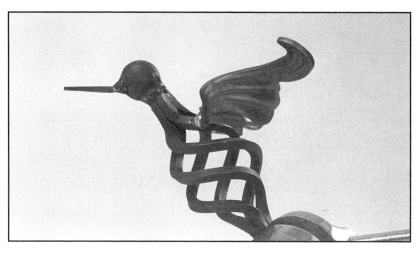

STEP 6-3 - *Using the pliers to hold the horseshoe nail, weld it to the steel ball.*

STEP 6-4 - *Place the 1/4-inch steel rod to the bottom of the bird's body and weld it.*

PROJECT 7: RUSTY, THE METAL COW

There are projects in this book that must be taken seriously, and then there are those whimsical projects that make you giggle. Rusty, the metal cow, is one of those. Use the body of Project 1 in this chapter.

Supply List

1 piece of 3/8″ steel plate 8″ wide × 10″ long

4 pieces of 3/4″ concrete reinforcing rod 1′ long

2 3/4″ flat washers

1 piece 1/4″ × 1″ × 6″ steel flat bar

2 horseshoes

1 piece of 1/8″ steel 12″ wide × 18″ long

1 piece of 1/4″ steel 15″ wide × 15″ long

Equipment Needed

a soapstone

a pair of safety glasses

a GMA welding machine

a welding hood

an oxyfuel cutting torch

a pair of cutting goggles

a torpedo level

a 4 1/2″ angle grinder

a sparklighter

a pair of leather gloves

Patterns

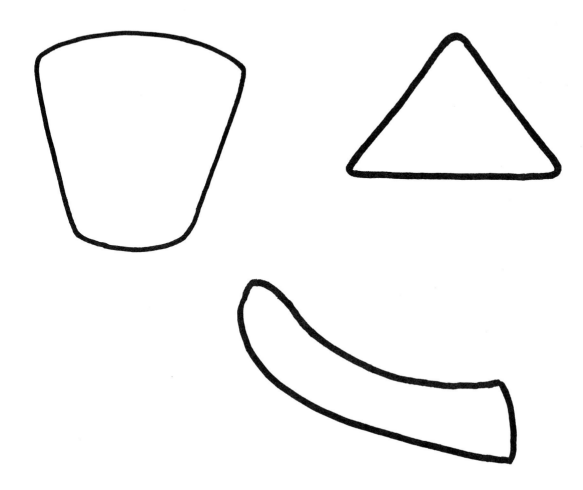

CHAPTER

6

CANDLE-HOLDERS

The goal of this chapter is to guide you, the artist and crafts-man, to produce a product, and have the freedom to experiment while having an enjoyable experience. Hopefully, it will influence you by combining contemporary and traditional objects with items you have found to create projects that are not of the norm. Let the unusual become the usual; try changing shapes, objects, dimensions, or angles. Try to create projects that will stimulate your creativity.

PROJECT 1: TOWERING CANDELABRA

You will probably want to make several of these. They look great in a cluster, particularly when you vary their heights.

Supply List

1 piece of 1/2″ square tubing 40″ long

1 4″ circle 1/4″ thick

3 Scrolls 1/4″ thick × 1″ wide × 7″ long

2 Scrolls 1/8″ thick × 1/2″ wide × 4″ long

1 5″ bobeche (the round circle that a candle sits on)

1 nail 1″ long

Equipment Needed

a GMA welding machine

a welding hood

a pair of safety glasses

a pair of leather gloves

a pair of pliers

a tape measure

***STEP 1-1** - Weld the 7-inch long scrolls to the 4-inch circle 120 degrees from each other.*

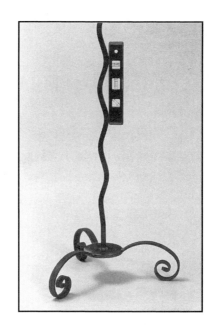

STEP 1-2 - *Weld the wavy 1/2-inch tubing to the center of the 4-inch circle.*

STEP 1-3 - *Weld the two 4-inch scrolls diagonally to each other and approximately 6 inches form each other. The first scroll should be 6 inches below the top of the 1/2-inch tubing.*

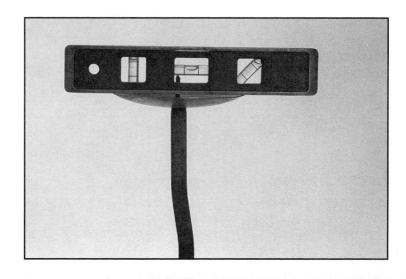

STEP 1-4 - *Weld the 5-inch bobeche with the 1-inch nail in the center to the top of the 1/2-inch tubing.*

PROJECT 2: ELEVATED CANDLEHOLDER

There is something different about the glow from the elevated candleholder. Could it be the use of geometrical shapes? The scale or proportion? Or the simple, yet, direct lines of the item? Sometimes it could just be the feelings of an object that draws your attention.

Supply List

2 pieces of 3/8" rod 20" long

5 pieces of 3/8" rod 3 1/2" long

2 pieces of 3/8" rod 28" long

2 pieces of 3/8" rod 8" long

2 pieces of 3/8" rod 3 9/16" long

1 piece of 16 gauge carbon steel 9" square

4 pieces of 1 1/2" pipe 1" long

4 5/8" washers

Equipment Needed

a GMA welding machine

a welding hood

a pair of safety glasses

a pair of leather gloves

a tape measure

a pair of pliers

a speed square

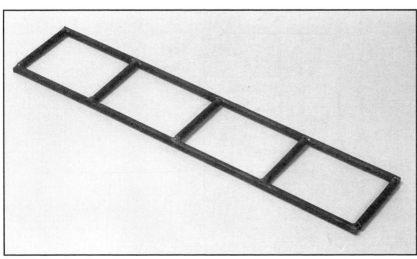

STEP 2-1 - *Weld together the two 20-inch pieces and five 3 1/2-inch pieces of the 3/8-inch bar. Make sure all piece are uniformly spaced and square to each other.*

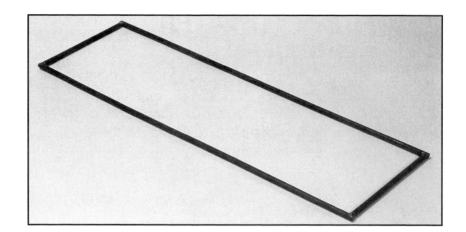

STEP 2-2 - *Using the two 28-inch pieces and the two 8-inch pieces, form a rectangle and weld the corners together.*

STEP 2-3 - *Place the small rectangle in the center of the larger rectangle and connect them with the two pieces of three 9/16-inch long bars.*

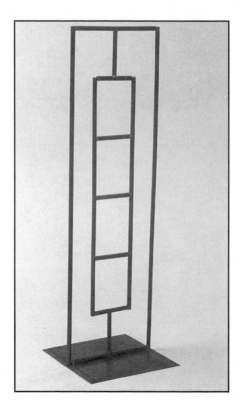

STEP 2-4 - *Center the weldment to the 9-inch square and weld.*

STEP 2-5 - *Weld each 5/8-inch washer to the bottom of each piece of 1 1/2-inch pipe.*

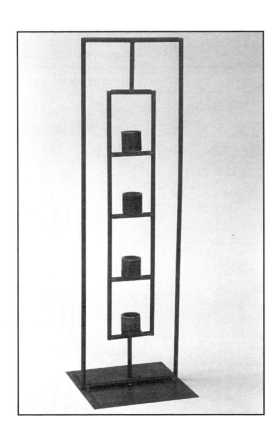

STEP 2-6 - *Center and weld each 1 1/2-inch pipe to each 3 1/2-inch piece of 3/8-inch bar.*

PROJECT 3: A CANDLEHOLDER SOMEWHERE BETWEEN THE ORDINARY AND THE EXTRAORDINARY

Supply List

1 20" circle made from 1" square tubing

2 pieces of 1/8" × 1" × 12" carbon steel

2 pieces of 1/8" × 1" × 16" carbon steel

2 pieces of 1/8" × 1" × 3 1/16" carbon steel

1 decorative casting approximately 5" in diameter

1 scroll 9" long

1 scroll 5" long

1 5" bobeche 1/16" thick

Equipment Needed

a GMA welding machine

a welding hood

a pair of safety glasses

a pair of leather gloves

a tape measure

STEP 3-1 - *Weld the 5-inch casting and 5-inch scroll to the 20-inch circle at the 12 o'clock and 6 o'clock positions.*

STEP 3-2 - *Weld both 12-inch and 16-inch pieces of 1/8-inch carbon steel into place.*

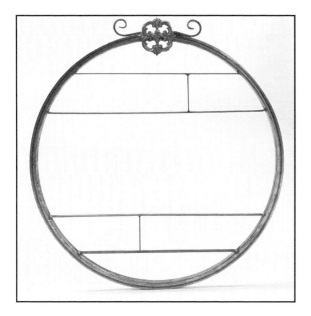

STEP 3-3 - *Randomly place the two 3 1/16-inch pieces.*

STEP 3-4 - *Weld the 9-inch scroll and the 5-inch bobeche onto the center of the bottom of the circle.*

PROJECT 4: PUMPKIN CANDLEHOLDER

This project is fresh, whimsical, and fun to make. It is a great project for the fall of the year.

Supply List

8-1/4 horseshoes

1 piece of 1/8″ rod 1 foot long

1 piece of 3″ × 3″ sheet steel

1 4″ bobeche

Equipment Needed

a GMA welding machine

a welding hood

a pair of safety glasses

a pair of leather gloves

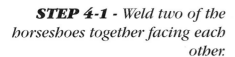

STEP 4-1 *- Weld two of the horseshoes together facing each other.*

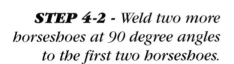

STEP 4-2 *- Weld two more horseshoes at 90 degree angles to the first two horseshoes.*

STEP 4-3 - *Place the 4-inch bobeche in the center of the four horseshoes and weld in place.*

STEP 4-4 - *Weld the remaining four horseshoes 45 degrees from each of the previous horseshoes.*

STEP 4-5 - *Weld the quarter horseshoe to the top center of the pumpkin.*

STEP 4-6 - *Coil the 1/8-inch rod around a piece of 1-inch pipe to make a coil. Weld the coil, along with the two leaves that you have fabricated form the 3-inch × 3-inch sheet steel to the top of the pumpkin.*

PROJECT 5: MUSIC STAND CANDELABRA

This project will serve as a stunning focal point in any room.

Supply List

5 pieces of 1/2″ square tubing 16″ long

2 pieces of 1/2″ square tubing 22″ long

9 4″ bobeches

1 9″ scroll

2 6″ scrolls

3 5″ scrolls

1 piece of 1/2″ twisted bar 24″ long

1 6″ circle 1/4″ thick

1 piece of 1/8″ × 3/4″ × 20″ carbon steel

Equipment Needed

a GMA welding machine

a welding hood

a pair of safety glasses

a pair of leather gloves

a tape measure

a torpedo level

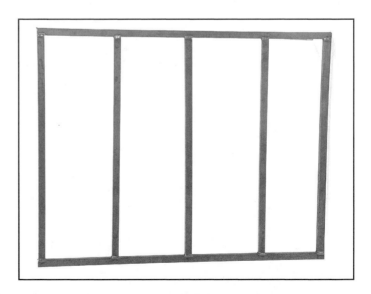

STEP 5-1 - *Using the five 16-inch pieces and two 22-inch pieces of 1/2-inch tubing, arrange and weld them into a four-section rectangle.*

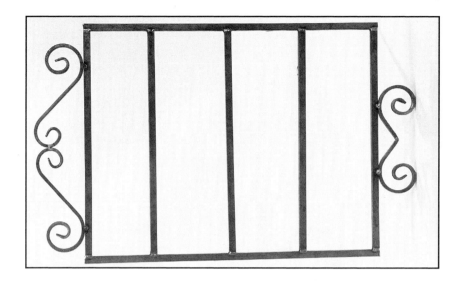

STEP 5-2 - *Weld the 9-inch scroll and the two 6-inch scrolls into place at the top and bottom of the large rectangle.*

STEP 5-3 - *Bow the piece of 1/8-inch × 3/4-inch × 20-inch carbon steel, and weld it to the center of the back side of the large rectangle.*

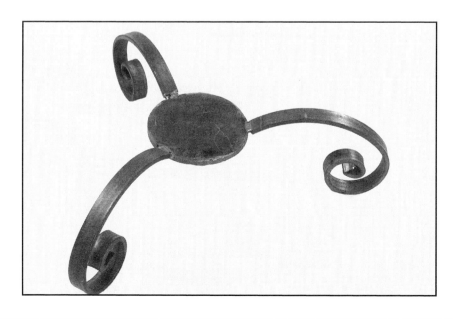

STEP 5-4 - *Weld the three 5-inch scrolls to the 6-inch circle 120 degrees from each other to form the base.*

STEP 5-5 - *Weld the 24-inch long twisted bar to the center of the 6-inch circle base.*

STEP 5-6 - *Attach the rectangular piece 45 degrees to the twisted bar by welding the bar to the bow on the back side of the rectangle.*

STEP 5-7 - *Center, level, and weld three bobeches on the three inside bars of the large recatangle.*

SOURCES

Company	Phone	Description
Akron Fitting Co.	330-753-0311	architectural metal components
Allen Robbins Arch. Metals, Inc.	205-234-3036	custom castings and patterns
Classic Iron Supply	800-367-2639	iron castings, forgings, and supplies
Crescent City Iron Supply, Inc.	800-535-9842	castings, spears, locks and hardware door castings, and paint
Doval Industries	800-237-0335	ornamental iron supplies
Eagle Bending Machines, Inc.	334-937-0947	pipe, bar and plate benders
Eagle Iron Supply, Inc.	972-289-7688	ornamental iron hardware
J. G. Braun Co.	800-323-4072	stamped steel ornaments, steel wood-grain rod and castings
Julius Blum & Co., Inc.	800-526-6293	architectural metal components
King Supply Co., Inc.	800-542-2379	iron and aluminum castings, spears and brackets
Lavi Industries	800-624-6225	architectural metal components
Lawler Foundry Corp.	800-624-9512	ornamental castings, forgings, and supplies
Omega Coating Corp.	888-386-6342	paint for ornamental iron
Russian Blacksmithing	011-70-74-277-5587	ornamental iron wares
Texas Metal Industries	800-222-6033	ornamental iron and aluminum castings, furniture, accessories, and commercial castings
Triple-S Steel Supply	800-231-1034	ornamental supplier
Universal Manufacturing Co., Inc.	800-821-1414	castings, locks, hinges and supplies
Wasatch Steel	801-486-4463	steel and supplies
Yavuz Ferforje Ve Demir Tic San	011-90-258-371-0923	decorative elements, twisted bars, scrolls, and rosettes